SpringerBriefs in Cybersecurity

Editor-in-Chief

Sandro Gaycken, Digital Society Institute, European School of Management and Technology (ESMT), Stuttgart, Baden-Württemberg, Germany

Series Editors

Sylvia Kierkegaard, International Association of IT Lawyers, Highfield, Southampton, UK

John Mallery, Computer Science and Artificial Intelligence, Massachusetts Institute of Technology, Cambridge, MA, USA

Steven J. Murdoch, University College London, London, UK

Kenneth Geers, Taras Shevchenko University, Kyiv, Kievs'ka, Ukraine

Michael Kasper, Department of Cyber-Physical Systems Security, Fraunhofer Institute SIT, Darmstadt, Hessen, Germany

Cybersecurity is a difficult and complex field. The technical, political and legal questions surrounding it are complicated, often stretching a spectrum of diverse technologies, varying legal bodies, different political ideas and responsibilities. Cybersecurity is intrinsically interdisciplinary, and most activities in one field immediately affect the others. Technologies and techniques, strategies and tactics, motives and ideologies, rules and laws, institutions and industries, power and money—all of these topics have a role to play in cybersecurity, and all of these are tightly interwoven.

The *SpringerBriefs in Cybersecurity* series is comprised of two types of briefs: topic- and country-specific briefs. Topic-specific briefs strive to provide a comprehensive coverage of the whole range of topics surrounding cybersecurity, combining whenever possible legal, ethical, social, political and technical issues. Authors with diverse backgrounds explain their motivation, their mindset, and their approach to the topic, to illuminate its theoretical foundations, the practical nuts and bolts and its past, present and future. Country-specific briefs cover national perceptions and strategies, with officials and national authorities explaining the background, the leading thoughts and interests behind the official statements, to foster a more informed international dialogue.

Laurie Yiu-Chung Lau

Cybercrime in Asia

Policing, Technological Environment,
and Cyber-Governance in China and Vietnam

 Springer

Laurie Yiu-Chung Lau
Asia Pacific Association of Technology
and Society (APATAS)
Shatin, New Territories, Hong Kong

ISSN 2193-973X ISSN 2193-9748 (electronic)
SpringerBriefs in Cybersecurity
ISBN 978-3-031-80212-6 ISBN 978-3-031-80213-3 (eBook)
https://doi.org/10.1007/978-3-031-80213-3

This Springer imprint is published by the registered company Springer Nature Switzerland AG
The registered company address is: Gewerbestrasse 11, 6330 Cham, Switzerland

If disposing of this product, please recycle the paper.

To my forebear's Lau Chiu Lan Tso, with him founded Tin Liu 'with his bare hand' the place called Tai Tung Wo Liu (Tin Liu) more than 400 years ago in the early eighteenth century AD, I would not have got the land that to build my house on today.

Preface and Acknowledgements

This book began as a personal motivation as well as an academic one. The idea could be dated back to 2018, about 10,728 km in distance away from Hong Kong at a historical city called 'Granada' in Spain. I had graduated and received my Ph.D. already for some time then, but at the back of my mind there was an ambition that I wanted to publish a book as a single author. In the past my scholarly works were published either as one of the chapters in a collection of edited books, which I often co-edited together with my peers, or articles in the international journal. On 17th of September 2018, I met Wayne Wheeler from the publisher of Springer Nature, who happened to be at the Granada Convention Centre to perform a formal business trade for a medical imaging international conference. During a casual exchange of conversations, I expressed to him the idea of writing a book on Cybercrime in Asia. I remember it clearly and it is still very fresh in my mind that Wayne's expressions were very positive and said to me that "Laurie, this is a bloody good idea." Armed with this idea of a book and after the conference was successfully completed, Wayne and I went separate ways. He went back to the UK, and I got back to my office in Hong Kong. We were exchanging a few emails to and fro with each other. Then this book idea was sealed with a formal contract signed in the middle of 2019 and the book was expected to be completed by the middle of 2020.

The topic of this book is also the subject area very much related to my scholarly works. More importantly, Asia is going to be a technological powerhouse for the coming twenty-first century, and leading the way is our mother country 'The People's Republic of China'. She is advancing rapidly in modern technologies, for works and for leisure. As Asia progresses towards advancing and modernising their technologies, it is going to be a hotspot for cybercrime. While Asian netizens potentially fall victims to cybercrime, they are also capable of becoming perpetrators of cybercrime too—carrying out the crimes domestically and internationally. Such double edge sword phenomenon of technological-related crime inspired me to write this book. It is hoped that this book could draw much needed attention on cybercrime in Asia—instead of global North, as technologies and crimes know no boundaries. And Asia can act early to prevent cybercrime before it is too late.

Writing this book is an arduous journey. Shortly after the book contract was signed COVID-19 brought a global pandemic lock down, cutting off the connection from the rest of the world here in Hong Kong for almost three and half years. It was only in late 2022 that Hong Kong began toe-trippingly opening up its border and international flight connections. In between these periods of time, my own father (Lau 'Yam' Sing) passed away from old age. Furthermore, as an independent scholar not holding a full-time position in academia, and as an entrepreneur running a business, I found myself multitasking on multiple things at the same time. Finding the time and sitting down quietly to write is very challenging indeed. All these obliged me to put this book on hold. It was not until late 2023 that I was slowly getting back to this book, putting my thinking cap on, and writing it.

However, this book will not be in the format as formerly planned back in early 2019. The original intention was a longer version with more chapters in one book, instead of a short adaptation of two chapters as Volume 1. It is expected that the book will be expanded at a later day in Volume 2 and Volume 3. In Volume 2 there will be more chapters on countries such as Indonesia, then the Philippines, followed by Thailand, and finally in Volume 2 the discussion will be turning to Singapore and Hong Kong, since the two places have been keen competitors for decades in trades and in economic developments. The two places, though politically and socially disparate, also share a lot of similarities in a number of ways. Singapore is a sovereign nation state while Hong Kong is a Special Administrative Region and part of the People's Republic of China since 1997. Singapore's population is also smaller than Hong Kong (5.6 million and 7.4 million, respectively). Geographically Singapore is located at the edge of the Malay peninsula and Hong Kong at the East of the Pearl River Delta. However, both of them were former British colonies and still adopting common law tradition as their legal system. Singapore is an international financial centre serving the Southeast Asia region and Hong Kong is also an international financial centre serving globally, as well as acting and bridging between mainland China and the world in recent years. Modern technological adoption and digitalisation are playing an important role in maintaining the financial centre status for both places. According to the Singapore Digital Economy Report 2023,[1] Singapore's digital economy is at 17.3% of Singapore's GDP in 2022, up from 13% of GDP in 2017. Whereas digitalisation among Small and Medium Enterprises (SMEs) had increased from 73.8% in 2018 to 94.3% in 2022, especially on e-payment systems. As for Hong Kong, the situation is a bit paradoxical. On the one side, as one of the key global international financial centres, Hong Kong enjoys a very high ranking by most global technological indices. For example, the coverage of 5th generation mobile network reached 90% of the Hong Kong population.[2] On the other side of the coin, Hong Kong's digital economy contribution to its GDP is lagging behind Singapore. In the

---

[1] See Infocomm Media development Authority and Lee Kuan Yew School of School Public Policy, National University of Singapore, 'Singapore Digital Economy Report 2023', <singapore-digital-economy-report-2023.pdf (imda.gov.sg)>, visited 14/8/2024.

[2] Innovation, Technology and Industry Bureau, the Government of the Hong Kong SAR, PRC, 'Hong Kong Innovation and Technology Development Blueprint', p. 53.

last survey by the Census and Statistics Department of the Hong Kong Government, in 2016, 5.7% of the digital economy attributed to the total Hong Kong's GDP.[3] This is partly due to the fact that Hong Kong did not consistently measure its digital economy, and only in 2023 the Hong Kong Government formally established a governmental office to measure digital economy systemically.[4]

Finally, in Volume 3 as I envisioned the attention and focus is on the analyses, put it all together, to offer some ideas on the way forward on fighting the ever-evolving cybercrime 'or technology-related crime' in Asia. By looking at the cybercrime problem from various lenses or perspectives: socio-legal, socio-economic, and equally important the geopolitics as all countries or city states covered in all volumes were being influenced by domestic, regional, and international geopolitics on a daily basis, leads to different responses. We are trying to identify any unique or difference so that we can learn from them. Perhaps in this final Volume it is intended to push the envelope further for a better understanding of cybercrime which currently stands. A novel idea applies ancient Greek philosophy to understand the phenomena of cybercrime. In the 1980s, scholars researching on computer-related crime looked at it through the lens of knowledge. However, from what we understand today, issues related to cybercrime are evolving constantly—as the societal factor changes so the impact on cybercrime is changing too. It therefore makes sense to understand it from another angle, from the perspective of knowledge of ignorance.

In writing this Volume 1, I incurred many debts to people who had been supporting and inspiring me. I am most grateful to Wayne Wheeler, who was with this book from the beginning, for his strong support in securing Springer Nature's continuous commitment and in realising my ambition of a book published as a single author, and to Sriram Srinivas of Springer Nature too for his constant reminders that it is time to complete and submit the manuscript for print—and I had done it. Last but not least, I want very much to thank my wife Lam Cheuk Man for her constant support and inspiration to complete this Volume 1 book. She helped me to do the very challenging task of proofreading and refining the first draft of this book. I also want to thank Momentous Asia Travel and Events funding my travel to places around the globe, from the South-pole to Europe. All these travels opened my eyes to the world and made me realise it is best to stay humble and to tie my feet to the ground firmly. Furthermore, I want to specially mention a person here, Mr. Lee Kwok Wah. He has inspired me for many decades now in different ways—not in scholarly works but through the other most important thing in life—the human skill that it is the practical societal skills and friendship. Similarly, I also want to mention a person who passed away in January 2003. His teachings are still very much inspiring me even today, I am in debt to my mentor at Wo Hing Construction Mr. Yeun Kam Pun. Scholarly, I am also indebted to Prof. Carol Jones, for her introduction to my

---

[3] See Census and Statistics Department, Hong Kong Government, (November 2018) 'Measurement of Digitalization in Hong Kong' <標題版面配置 (stats.gov.cn)>, visited 15/8/2024.

[4] See Innovation, Technology and Industry Bureau, the Government of the Hong Kong SAR, PRC, 'Hong Kong Innovation and Technology Development Blueprint'.

understanding of social-legal studies. Finally, I want to thank all those people who I forgot to mention here and who are helping me along the way too. Nonetheless, as ever, in this Volume 1 all the mistakes and obligations remain to be mine.

Shatin, Hong Kong   Laurie Yiu-Chung Lau

Contents

Chapter 1
Cybercrime in Asia and Its Development

1.1 Introduction

To start the book, it is better for us to get some understanding about Asia. Firstly, there is a nation block called ASEAN. It is an establishment of Association of Southeast Asian Nations (ASEAN) block at the Bangkok Declaration in August 1967. The main aims of founding of ASEAN were cooperation in the social, economic, cultural, technical, educational, and so forth, from very fragmented and developing nations in the Southeast Asia region which even today are still very fragmented. This group of nations is complex in terms of their economic developments, comprising a mix of developed and underdeveloped countries, particularly if looking from a Western liberal democracy perspective. Only a few nations within the ASEAN are observing a one-man-one-vote process to elect their national leaders. The majority of them are in between semi-authoritarian to authoritarian states. For example, Singapore is technically democratic in the ballot box process, but in practice is subject to interpretation.

Today, ASEAN has a population of more than 5 billion. The rapid advancement of modern information technology and telecommunication technologies within the block has directly and indirectly transformed and benefited ASEAN's populations. For example, rural areas can now access banking services via mobile Internet technology, 24/7 shopping on the mobile e-store apps, etc.....However, the nature of these modern technologies has positive and negative impact on people's daily life. For instance, in areas of AI, robotics in medicine, such as nano and biotechnologies. These technological advancements could be a double edge sword: they could cure disease and better or improve quality of life while at the same time the uneven spread of such technologies in ASEAN nations is opening the gap between so-called 'haves' and 'have-nots'. The 'Have-nots' are more likely to fall victims of cybercrimes, due to lack of technology awareness and education. According to the World Economic

L. Yiu-Chung Lau, *Cybercrime in Asia*,
SpringerBriefs in Cybersecurity, https://doi.org/10.1007/978-3-031-80213-3_1

Forum 2023 report, the Asia Pacific region including the ASEAN is the prime target for cybercriminals to pry their trade on as the diagram[1] shown below.

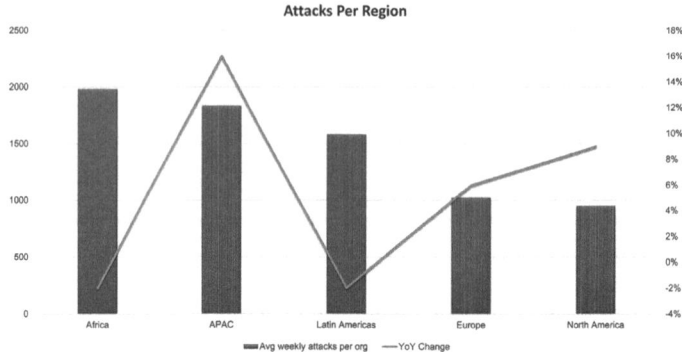

As Asia is a big region, discussion in this first Volume 1 will largely be confined to a few Asian countries. Asia region is relatively large and geographically it covers an area of 44 million square kilometres, where 60 percent of the total world population is residing. This population is ethically and culturally diverse. There is no common language and religion in these 55 nation states (including city state). This would be a task too big for one book volume to tackle. To this end, the author would identify a few best fit nation states or city states such as Vietnam and the People's Republic of China (PRC); look at how cybercrime stacks up in these countries for a good snapshot on cybercrime in the region; and explore issues such as does economic development make a difference on capability and resource allocation in fighting cybercrime, or less economically developed nation state attracts more cybercrime, and if this is the case, why so as the general assumption is that cybercrime is boundless and every nation state should be facing the same level of threat of cybercrime.

1.2   Digitalisation in Asia

In recent years, Asian region countries are the new 'ground zero' for cybercrime. According to a report by Check Point Research,[2] Southeast Asia countries witnessed the highest year-over-year increase in weekly cyberattacks during the first quarter of 2023, averaging 1,835 attacks per organisation. In contrast, the global average stood

---

[1] See World Economic Forum <Asia Pacific region is the new ground zero for cybercrime | World Economic Forum (weforum.org)> . Accessed 3/11/2023.

[2] See Global Cyberattacks Continue to Rise with Africa and APAC Suffering Most - Check Point Blog. Accessed 3/03/2024.

at 1,248 attacks per week. A report by the Bangkok Post newspaper[3] on 19th June 2024 described that by 2025, cybercrime loss is estimated to value at United States Dollar (US$) 10.5 trillion, globally up from US$3 trillion in 2015 from a global perspective. Such an alarming upward trend does beg the question of what is behind this shift. To answer this question, there are several factors or trends likely at play here, not just in ASEAN, but globally too. For example, on 28 August 2024, a BBC news report on a Nigerian Cybercrime Network called 'Black Axe'. Its members were well educated with university degrees. They were laundering hundreds of million US Dollars from cyberscams, frauds, and their operations were global. As Interpol said:

> "Black Axe members are university educated and are recruited during their schooling, its organization is a secretive criminal network…operating around the world…Bank robberies are now done with laptops - they're far more sophisticated…cryptocurrency - which can be sent rapidly between digital wallets around the world - is becoming an integral element in Black Axe's money-laundering operations."[4]

Here are some of the factors:

Accelerated digital transformation—Asian countries have undergone rapid digital transformation, especially during and post COVID-19 pandemic. Many organisations rushed to adopt new technologies and digital platforms, often without sufficient security measures installed making them extremely vulnerable, and opening opportunities for exploitation. This increased digitalisation offered a seedbed for cybercriminals.

A new generation of users—TikTok and Facebook generation in ASEAN countries rely heavily on mobile devices and collaborative tools, to the point they have become desensitised to the risks associated with clicking on suspicious links or sharing sensitive information online, combined with lack of education on public awareness on digital technologies. As a result, their online habits make them more susceptible to social engineering attacks and phishing attempts.

Hybrid working mode—The growing popularity of hybrid working arrangements during and after COVID-19 lockdown period led to increased reliance on digital communication and collaboration tools, exposing organisations to new security risks. Cybercriminals exploit vulnerabilities in remote access systems and unsecured access points within home.

---

[3] < See Bangkok Post, <https://www.bangkokpost.com/business/general/2813479/fortinet-points-to-surge-in-value-of-cybercrime-by-2025>. Accessed 19/06/2024.

[4] See BBC Online (28 August 2024) 'World's police in technological arms race with Nigerian mafia' < Nigeria's Black Axe mafia dealt 'big blow' by Interpol (bbc.com) > . Accessed 28/08/2024.

The collaboration riddle—The booming of collaboration application platforms opened a new pathway for cybercriminals. Increased usage of applications such as video conferencing, cloud storage, and file-sharing platforms gives rise to potential security breaches, with potential perpetrator targeting weak security settings, unpatched software, and unsuspecting users to gain unauthorised access to sensitive data.

Asia Manufacturing Hub—Asian countries like Taiwan, China, and Vietnam are electronic manufacturing hubs, in particular, over semiconductor and manufacturing. The economic importance and intellectual property these industries hold make it an attractive target for cyber espionage and intellectual property theft.

Today, Asian nations have the political aspiration by leveraging the information technologies as their driving forces for economy growth. Bangkok Post reported that the Deputy Prime Minister of Thailand made a speech on 11 November 2020 at The Powering Digital Thailand 2021 conference in Bangkok. The reporter Gen Prawit said:

> Thailand is one of the first countries in Asean that developed 5G technology for commercial use. Many areas in the Eastern Special Development Zone or Eastern Economic Corridor (EEC) have already adopted this technology. Digital technologies like 5G, Cloud and AI technologies are crucial infrastructure that can turn Thailand into the digital hub of the region"[5]

As he went on and said:

> The government [Thailand] recognised the importance of technological development…It therefore drew up its Digital Economy and Society Development Plan in cooperation with the public and private sectors, he said. Its aim was to develop infrastructure, innovation, data, human capital and other digital resources that will ultimately drive the country towards stability, wealth and sustainability.

1.3 Challenges

The more the nation is dependent on the information technology for their economic growth, the more likely for perpetrator to pry on their netizens and population. This is the other side of the digital economy coin. Table Chart[6] below shows the percentage of online users who were victims of cybercrime from October 2016 to September 2017. It tells how widespread cybercrime was and that certain nations were more prone to it than others: China, India, Indonesia, and United States were high as victims

---

[5] The Bangkok Post <https://www.bangkokpost.com/business/2017999/digital-hub-status-within-reach>. Accessed 12/11/2020.

[6] See SCMP <https://www.scmp.com/news/hong-kong/community/article/2139170/least-2-million-internet-users-hong-kong-were-hit>. Accessed 27/02/2020.

of cybercrime, while Hong Kong was relatively low in terms of Cybercrime level. Back then in Hong Kong in 2017, usage of mobile payment and mobile application among its netizen was not as popular.

Global impact of cybercrime

The percentage of internet users who were victims of online crimes from October 2016 to September 2017

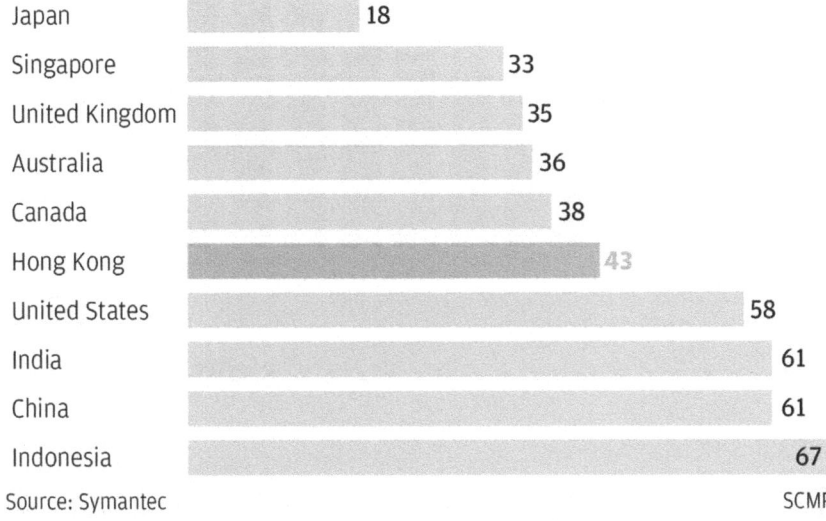

Japan	18
Singapore	33
United Kingdom	35
Australia	36
Canada	38
Hong Kong	43
United States	58
India	61
China	61
Indonesia	67

Source: Symantec   SCMP

Globally cybercrime is rapidly increasing and naturally the loss induced by cybercrime will increase too. The Table below shows year by year increment, and by 2028, it is expected that the loss incurred by cybercrime will reach 13.82 trillion USD.[7] It was only 0.86 trillion USD ten years ago in 2018, a 13-fold increase in 10 years.

---

[7] See Statista < Chart: Cybercrime Expected To Skyrocket in Coming Years | Statista>. Accessed 24/07/2024.

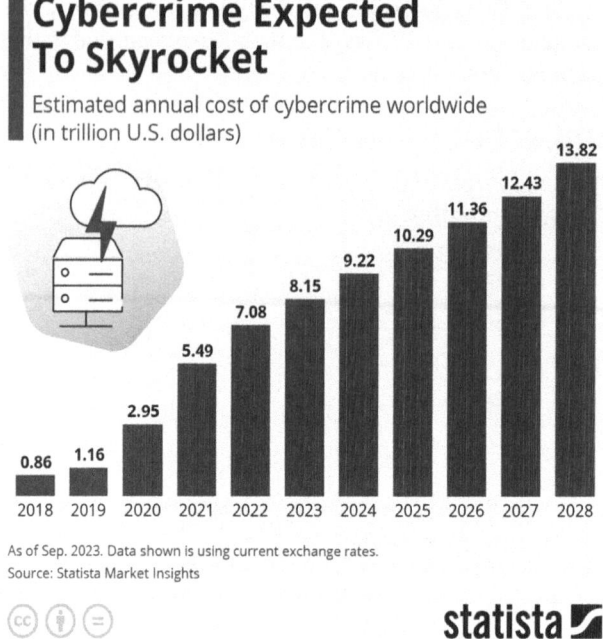

Cybercrime Expected To Skyrocket

Estimated annual cost of cybercrime worldwide (in trillion U.S. dollars)

As of Sep. 2023. Data shown is using current exchange rates.
Source: Statista Market Insights

statista

1.4 Effect to Stamping the Tide of Cybercrime in Asia

As resources are finite, different sovereign nations would handle it differently to stamp the tide of rise in cybercrime within their national border, and measures taken are largely according to national income and wealth. As the majority of these Asian countries are economically developing nations, it is going to be a challenge in terms of capability and resource building. As a result, in recent years, with the spikes of cybercrime in the Asia region, some Asian countries such as Malaysia, Thailand, and Vietnam are more prone to cybercrime than others caused by a sudden acceleration of the general public adopting information technologies and mobile-phone applications. Other economically developed countries or city states are better equipped to fight the rise of cybercrime by investing in technologies to fight technology scams. Singapore and PRC were among the first Asian countries in the world to officially use facial verification in their national identity scheme, by using cloud-based face verification technologies. Such biometric check schemes allow Singaporeans to secure access to both private (i.e. banking services) and government services. As the government's technology agency,[8] Mr. Bud says:

> it will be "fundamental" to the country's digital economy. It has been trialled with a bank and is now being rolled out nationwide. It not only identifies a person but ensures they are genuinely present...You have to make sure that the person is genuinely present when they authenticate, that you're not looking at a photograph or a video or a replayed recording or a deepfake," said Andrew Bud, founder and chief executive of iProov, the UK company that is providing the technology. The technology will be integrated with the country's digital identity scheme SingPass and allows access to government services.

And in Hong Kong too, the Hong Kong Police is using artificial intelligence tools to fight the rising tide of ransomware cases, love scams, investment frauds, phishing schemes, and identity thefts, as reported by SCMP Newspaper[9]:

> ...The increase in the prevalence of cybercrime in Hong Kong has been dramatic and it continues to accelerate. There were more than 22,000 cases last year involving losses of HK$3.2 billion, a fourfold increase over the last six years. The trend has continued in 2023, with 7,301 cases in the first three months, 59 per cent more than the same period in 2022. The crimes are becoming increasingly sophisticated...

Indeed, it is clear that the governments of the wealthier nations or city states in Asia could channel budget and human resources to try to stamp the rise of cybercrime. Nevertheless, cybercrime problems are never confined to a certain geographical nation border, as the Internet network is global and boundless, which makes cybercrime a global issue.

In view of the above, this volume is unfolded as follows: after the introductory chapter, Chapter Two will explore how nations such as Vietnam are dealing with the challenges entailed by cybercrimes. Vietnam has been rapidly growing economically in recent years, and technologisation is one of the main strives behind such economic growth. Technology transformation and digital economy in Vietnam are key drivers because the Vietnamese government wants to shift its socio-economic growth model from a low technology-based to a digitalised-based model, as a means to closing the economic gap with developed countries. Vietnam is aiming 30 percent of Gross Domestic Product (GDP) of its economy will be generated from digital transformation within the society by 2030, while back in 2007 it was only a mere 7 percent[10] This rapid digital growth in the economy would create both positive and negative spillover effect, and undoubtedly is a disruptive force towards Vietnamese society. In Chapter Three the focus is on The People's Republic of China (PRC), as China is

---

[8] See BBC Online Report 'This is the first time that cloud-based face verification has been used to secure the identity of people who are using a national digital identity scheme' < https://www.bbc.com/news/business-54266602 > . Accessed 28/10/2020.

[9] See SCMP Newspaper Editorial 'Hong Kong's new cybercrime team can help turn rising tide of scams' 15 July 2023, < Opinion | Hong Kong's new cybercrime team can help turn rising tide of scams | South China Morning Post (scmp.com) > . Accessed 1/08/2024.

[10] See Dang Thi Viet Duc et al., 'Measuring the digital economy in Vietnam', in Telecommunication Policy, Vol. 48, Issue 3, April 2024, https://doi.org/10.1016/j.telpol.2023.102683. Accessed 12/08/2024.

at transitional period from the mid-level of developing nation to a higher economic nation state. It is expected to be a global economic powerhouse by 2035. No surprise to anyone that in recent years China's digital economy growth has been incredibly impressive achieving ahead of target by 2 years. China is now aiming for digital economy industries to contribute 10 percent to its GDP by 2025, up from 7.8 percent in 2020 which is laid down by the State Council's 14th Five-Year Plan period (2021–2025).[11] Back in 2019, 30 percent of its GDP is generated by the digital economy[12], by 2022 it had reached 41.5 percent, which made China rank second globally—in US dollar term worth 50.2 trillion Chinese yuan (USD7 trillion)[13].

1.5 Summary

This short introductory chapter highlighted that economic and social developments in Asia are diverse in nature. Some countries are more less on par with global north countries and have reached the level classified as 'developed economy', while most other Asian countries are at their developing stage and very few countries may have reached the middle level of their economic development. However, if not all, most countries in Asia have been leveraging technologies as strategies and toolkits for their nations' economic advancement.

Nevertheless, technology is a double-edged sword. On one hand, it attracts digital economic investment and boosts the nation's technological manufacturing hub, therefore, quickening the pace of digitalisation within the country. On the other hand, too fast a pace of digitalisation within the population could cause negative side-effect in the society. As we saw during the COVID-19 global pandemic, the unexpected rush into digitalisation globally has seen the level of cybercrime increased by 2 to 3 times as compared to normal trend[14]. Therefore, as Asia becomes more digitalised in their societies, the Asia region would become the prime target for cybercrime, which will be an impact and a directly challenge—at least in terms of their capability and finance resources to fight cybercrime. Next Chapter is turning to Vietnam.

---

[11] See The People's Republic of China, State Council, Press Release in English, < Digital economy expands in scale, demonstrating enormous potential (www.gov.cn) > . Accessed 14/08/2024.

[12] See International Monetary Fund Working Paper, Longmei Zhang, Sally Chen (2019) 'China's Digital Economy: opportunities and risks', p.4.

[13] See The People's Republic of China, State Council, Press Release in English, < Digital sector roadmap to aid recovery (www.gov.cn) > . Accessed 14/08/2024.

[14] See R. G. Smith, R. Sarre, L. Y-C. Chang, L. Y-C. Lau (Eds.), (2023) *Cybercrime in the pandemic digital age and beyond. Palgrave studies in cybercrime and cybersecurity.* Palgrave Macmillan: London.

Chapter 2
Vietnam

2.1 Introduction and Context: Vietnam

Vietnam is a Socialist Republic led by the Communist Party of Vietnam (CPV). The CPV follows the political philosophy and ideology of the late Hồ Chí Minh. Government powers in Vietnam are divided into legislative, executive, and judiciary functions—but collectively four members of CPV hold all powers in governing Vietnam, namely: General Secretary of the CPV, President of Vietnam, Prime Minister of Vietnam, and Chairman of the National Assembly of Vietnam. Unlike Common Law legality, Vietnam's legal system is based upon socialist legality. In common law tradition influential rule-making is the role of courts, whereas courts in socialist legal system play a dependent role. More importantly, although individual citizens do retain certain procedural and statutory rights in a socialist state, its citizens are expected to be obedient to the state—a situation similar to the People's Republic of China which will be further discussed in Chapter Three. Vietnam with a population of over 100 million was finally unified in 1975 after a long bitter war with colonial powers. Today Vietnam is one of the fastest growing economies in Southeast Asia, with year-on-year GDP growth reaching close to 8% with an estimated per capita income of USD 8,100.[1]

Vietnam implemented economic reforms in 1980 and 1990s. By 2020s, despite COVID-19 global pandemic and the increasing trend in deglobalisation, Vietnam still managed to become Asia's top-performing economy. Vietnam now manufactures higher-value goods and jobs are better paid due to its more highly skilled workers. These workers are now producing electronics which make up 38% of Vietnam's exports (compared to 14% in 2010). Vietnam had achieved an average of 6.2% in economic growth (faster than any other countries in Asia after China). Its

---

[1] International Monetary Fund Estimate for 2019.

growth is largely related to information technology sector, especially software development outsourcing export. Since the mid-2000s, Vietnam's communist government introduced and opened its market to trade: manufacturing and as information technology were identified as key drivers to drive economic growth. Today we can clearly see Vietnam high-tech industries forming a large and fast-growing part of the national economy and market-oriented trade policy rooted further into their economy. Although a latecomer to high tech, its adoption rate on information communication and technology 'ICT' is agile. The Ministry of Information Communication Technology published the Vietnam ICT White Book in 2014; by 2010, the total number of Internet users in Vietnam reached 26.8 million with a growth rate of 31% (Ministry of Information and Communications 2014). The increase in the number of Internet users in Vietnam can be seen year after year—from 31.3 million in 2011, to 35.6 million in 2012, to 40.1 million 2013, to 44.6 million in 2014, to 47.5 million in 2015, and in 2016[2] it reached 49.06 million. Another example was the smart mobile-phone took up rate in Vietnam. In 2017, 29% of the population, or '28.77 m' subscribed to smartphone. By 2022 the estimated number of smart mobile-phone users reached 42.66 million, more than 42% of the population. This also explains why Vietnam has become the prime target for cybercrime.

2.2   Cybercrime in Vietnam and Its Criminal Justice System

Vietnam's economy growth and its further opening of Internet connection with the rest of the world also created opportunity for cybercrime conducted domestically and trans-nationally. With a population of just over 100 million, of which 66% are engaged in Internet usage and 60% are mobile social media users, Vietnam is a country with rapid economic and digital growth, and in the midst of mobile social media lift-off too. It has seen an increment of 16% in its social media usage since 2018, of which usage on Smartphones is 72%, on laptop/desktop 43% among the adult population. On average, Internet access per day on any device reaches 6 h 42 min. As KrAsia[3] puts it:

> where there is big [economy] growth, there are big opportunities for cybercrime. Unfortunately, Vietnam has been ranked as the third worst in the world for its cybersecurity, behind Algeria and Indonesia, ranked at positions one and two respectively for their poor scores.

Then KrAsia went on and said:

> [Vietnam] rapid economic and digital expansion can often create a time lag in the adoption or dissemination of preventative measures to protect individuals [on cybercrime] and the wider national or global community as the region's digital growth exceeds expectations.[4]

---

[2] See: http://www.internetlivestats.com/internet-users-by-country/. Accessed 11/03/ 2020.

[3] See: https://kr-asia.com/vietnam-suffers-the-most-southeast-asia-offline-cyber-attacks-q2-2019/. Accessed 17/03/2020.

[4] Op Cit.

One classic example of cybercrime was in 2016: on 29 July two Vietnamese international airports, the Tan Son Nhat International Airport and Noi Bai International Airport, were hacked. It was alleged that a group, suspected of coming from China, launched hacker attacks on the website of Vietnam Airlines. Client information leaked and flight information screens at the two airports were disrupted, rendering the check-in computer system stop working for more than 60 min, causing flight delays. Altogether Noi Bai Airport has 30 flights, and Tan Son Nhat has more than 60 flights delayed, affecting about 2.000 passengers. Another cybercrime example is a home-grown Vietnamese hacking group called APT32, allegedly directly connected to the Vietnamese government, which used increasingly sophisticated cyber-attacks to spy on competitors and help Vietnam to catch up with global competitors. From the beginning of 2019 the APT32 group was actively hacking into and created fake domains for Toyota Motor Corp. and Hyundai Motor Co. in an attempt to infiltrate the automakers' networks. In March 2019, Toyota discovered that they were being targeted via Vietnam and Thailand and through a subsidiary—Toyota Tokyo Sales Holdings Inc—in Japan.[5]

2.3 Criminal Justice System

There are three establishments in Vietnam's criminal justice system to combat crimes, including cybercrimes: the police in charge of investigative process, prosecutor for prosecution, and the courts supposedly to independently obey only the law, but in practice may not always be the case. As Le, L. C., Hoang, Y. H., Bui, H. T. et al.[6] put it:

> In Vietnam, more than 98% of crimes are investigated by the police's investigative agencies. While the procuracies exercise the right to prosecute and oversee judicial activities, the courts shall exercise judicial power. Although the Constitution requires judges and people's assessors to conduct trials independently and obey only the law, this principle has not reached the full level of judicial independence in practice.

This is likely to have an impact on the efficiency of the criminal justice system, which may open up corruption practices in the justice system. According to Transparent International Corruption Perceptions Index 2019,[7] Vietnam scored 37 out of 100 points. Its ranking is amongst the two-thirds of the world's countries whose score was below 50, an indication that corruption in the public sector is a serious problem in Vietnam.

---

[5] See Bloomberg: https://www.bloomberg.com/news/articles/2019-12-22/vietnam-linked-hacking-group-targets-toyota-other-companies. Accessed 25/03/2020.

[6] See Le, L.C., Hoang, Y.H., Bui, H.T. et al. Understanding Causes for Wrongful Convictions in Vietnam: a View from the Top and the Bottom of the Iceberg. *Asian J Criminol* **17** (Suppl 1), 55–73 (2022). https://doi.org/10.1007/s11417-022-09390-7.

[7] See Transparent International: https://towardstransparency.vn/en/vietnam-cpi-2019-score-is-up-but-corruption-remains-serious/. Accessed 17/07/2020.

2.4 Response to Cybercrime

Early regulatory responses to technology-related crimes were very loosely put by the Vietnamese's communist government, as initially they were a late comer on information technology and Internet services became available to the general public only in the past decades. In the late 1990s Internet-related crimes started to emerge. In response, the Vietnamese National Assembly passed a law and for the first time codified a criminal law on cybercrime. However, such new criminal law on cybercrime was directly referenced from Russian laws and only criminalised a few Internet-related crimes. Under such background, the Criminal Code of Vietnam 1999 (CCV 1999) lacked precision and was very vague on the law[8]. Therefore, for almost 10 years there were hardly any Internet-related crimes being prosecuted by the authorities after the law was passed by the Assembly. Then in mid-2009, the Vietnamese communist government revised the 1999 CCV, refining the law by injecting more objectivity into the code, including crime against computer networks, Internet, digital devices, etc.... and at the same time, adding two more new offences related to cybercrime. Despite the revision of the CCV 1999 a serious loophole was still left open in the cybercrime legislation, as Luong et al. pointed out that: "...the disparities were confirmed as many Vietnamese criminologists and law enforcement officer stated that the implementation of [*the 1999 CCV*] had such serious loopholes..." For example, there were no spreading virus and denial of service (DoS) prosecuted under the 1999 CCV, because, as Luong et al. said "...[under 1999 CCV] considered the seriousness of damages caused by cybercrime as a compulsory factor to accuse criminals, but the investigation agencies struggled to detect the exact costs of the crimes. As a result, a large number of cybercriminals fell outside the scope of prosecution."[9] Then in 2015, to resolve the difficulties in the 1999 CCV provisions the Vietnamese National Assembly passed the Criminal Code 2015 for cybercrime (CCV 2015) and repealed the 1999 CCV. Nevertheless, before the CCV 2015 came into effect, the Vietnamese Communist government discovered several flaws in the provisions and the CCV 2015 was sent back to the National Assembly for revision. Eventually the law did not become effective until mid-June 2017. The CCV 2015 provisions consist of two parts, namely general provisions for definitions and the criminal part, laid out in chapter objectively of criminal offences. It expressively stated in the code that which offence is punishable under and to what type of criminal object that is upheld by the Vietnamese Constitution and Laws. The CCV 2015 is a big breakthrough from the 1999 CCV in the regulatory response to cybercrime, but whether this new provision is robust enough to deal with the fast-growing threat of cybercrime in Vietnam is still questionable.[10]

---

[8] Luong et al. 'Understanding Cybercrimes in Vietnam: From Leading-Point Provisions to Legislation System and Law Enforcement' K. Jaishankar: Open Access, International Journal of Cyber Criminology July–December 2019 Vol. 13, p. 299.

[9] Op Cit. p. 300.

[10] Op Cit. p. 301.

2.5 Police Response

Law enforcement agencies are playing a key role in combating cybercrime, and Vietnam is no exception, Vietnamese Police under the Ministry of Public Security established a Division of Cybersecurity and High-Tech Crime Prevention and Control. This Division is the sole agent for investigating cybercrime within its border. Some reports suggested that Vietnam is one of the top ten countries in the world for both cybercrime victims and perpetrators committing cybercrime. For example, in 2024 first quarter alone, the volume of cybercrime has been increased to 20 million, up from 130,000 cyber-attacks in 2017.[11] As some critics suggested, Vietnam authorities including the police who were struggling to contain growing cybercrime ecosystem,[12] the Vietnamese cybercriminals are capable to be global players, which would not be a surprise. According to GlobalEconomy[13] in 2018 Vietnam literacy rate was high, reaching 95%. Yet Vietnam funding in higher education is the lowest at 0.25% of its GDP when compared to other countries in Asia such as Thailand, Malaysia, and Indonesia, which account for 1% of their GDP. However, Vietnam public fund for general education, including mathematics and sciences, is relatively high at 80% of its overall expenditure. Although, its unemployment is low at 2%, according to QSWowNews[14] report the graduate unemployment rate is exceptionally high at 17%. There are a number of reasons behind, and one of them is the lack of relevant skills as university teaching is based on rote learning and the main purpose of studying is to pass exams, while many private and foreign companies are demanding more highly skilled workers. At high school level, there is, in a long tradition, a strong focus on science and mathematics, and such students frequently won awards at international science competitions. Therefore, there is a large pool of population skilled in IT but lack employment opportunities, while corruption is high and policing capacity on cybercrime is low, as well as support from existing criminal justice networks being also low, such as overall financial support on policing cybercrime is low too, as Vietnam relies heavily on private organisations to do the policing on cybercrime.[15] All these factors help make Vietnam a cybercrime regional hub in Southeast Asia,[16]

---

[11] See Vietnam + Sci-Tech April 4, 2024 'Over 20 million cyberattack warnings detected in Q1', < Over 20 million cyberattack warnings detected in Q1 | Vietnam + (VietnamPlus),. Accessed 23/08/2024.

[12] See Bank Info Security, Jayant Chakravarti (April 9 2024) 'Vietnam Struggling to Contain Growing Cybercrime Ecosystem', <Vietnam Struggling to Contain Growing Cybercrime Ecosystem (bankinfosecurity.asia)> . Accessed 23/08/2024.

[13] See GlobalEconomy: https://www.theglobaleconomy.com/Vietnam/Literacy_rate/ Accessed: 29/07/2020.

[14] See QSWowNews: https://qswownews.com/university-education-may-lead-worse-job-prospects-vietnam/. Accessed: 29/07/2020.

[15] See Bank Info Security, Jayant Chakravarti (April 9 2024) 'Vietnam Struggling to Contain Growing Cybercrime Ecosystem', <Vietnam Struggling to Contain Growing Cybercrime Ecosystem (bankinfosecurity.asia)> . Accessed 23/08/2024.

[16] See Lusthaus, J 'Modelling cybercrime development: The case of Vietnam' in 2020 (eds.) LeuFeldt R Holt T J, The Human Factor of Cybercrime, London: Routledge, p. 246–55.

a reflection of ineffective response to cybercrime by the authorities, including the police.

2.6 Summary

In summing up, Vietnam is a one-party state, and in recent years, we have seen rapid growth in its economy, in particular in information technology sectors. Yet despite its GDP growth the authorities are seemingly struggling to get a grip on the fast rate of growth of cybercrime. As a relatively underdeveloped country in Southsea Asia, such GDP gain does not translate into policing capability. Dependence on private organisations to police cybercrime proves to be insufficient and ineffective. Therefore, the gap between policing and governance and cyber-criminal both from within the country and outside of Vietnam border is very wide indeed. In next chapter we will be turning our focus on the People's Republic of China (PRC).

Chapter 3
The People's Republic of China (PRC)

3.1 Introduction and Context: PRC

The People's Republic of China is the third-largest country in the world in terms of land area, with a total land mass of approximately 9,600,000 km^2 (with 23 provinces, five Autonomous regions, four direct administered municipalities collectively is called as 'Mainland China', and two Special Administrative Regions of Hong Kong and Macau). Its population is approximately 1.4 billion. All these make China a mega-diverse and complex governance, especially in politics. Politically, modern PRC, like Vietnam, is under a one-party leadership, but China is vastly different in many ways from Vietnam. Since 1978, under the leadership of Deng Xiao Ping with the introduction to open door policy, China's governance has been reformed. Although politically centralised, economically is largely decentralised, of which today it is called 'capitalism with Chinese socialist character'.

Fast forward to twenty- first century China is economically a super powerhouse in Asia and globally. Since its reform in 1978, China super-leaps into a major player in global politics and economy. There is a saying that 'if China sneezes the world catch a cold'. By 2022, China is the second largest global economy in terms of GDP and the world's largest in terms of purchasing power parity (PPP), reaching 17.9 trillion US Dollar (approximately hold 18 per cent of global wealth). China today is ranked as an upper-middle-income country by the World Bank. Furthermore, by 2024, its economy is very diversified including manufacturing, retail, mining, Eclectic automobiles and energy generation, banking, electronics … notably information technology (with two-thirds of world patents on robotics inside China), and E-commerce.

In fact, China since ancient time has been an innovator in both science and technology. During Song Imperial Dynasty (late nine centuries) papermaking, printing, compass, and gunpower were already widely used. It was only in seventeenth century Western power surpassed China and for almost two centuries China had been weakened at a result of Western colonial powers and Imperial Japan invasion. When the Communist Party came to power in 1949, the country's science and technology were

L. Yiu-Chung Lau, *Cybercrime in Asia*,
SpringerBriefs in Cybersecurity, https://doi.org/10.1007/978-3-031-80213-3_3

gradually reviving systemically based on Soviet model. By 2023 China ranked 12th in the Global Innovation index and Chinese supercomputers ranked among the fastest in the world. These achievements were no accident. China believe strongly in education, with a strong emphasis on science, technology, engineering, and mathematics educations. In 2020 3.57 m graduates with STEM in China (40 percent of the total graduates), ranking top position in the world.[1]

On March 3rd, 1986, the PRC government intended to be decoupling from and not to be dependent for foreign technologies. She launched a strategic advancement plan in all fields in technologies called '863 plan' to nurturing home-grown technologies. By 2020s China frog-leaps into a new era in the realms of advance technologies and becomes a true techno-nation, which is almost unrecognisable to many. According to World Intellectual Property Organization (WIPO) in 2017 China has already surpassed the United State for filed patent applications. In 2023 almost 15 percent of the total Chinese population in China was scientific literary as surveyed by the China Association for Science and Technology[2] In terms of technologies in day-to-day applications, China is the largest e-commerce market in the world, with a total of 1.4 billion mobile subscribers (and mostly on 5G network) which China 5G mobile network has covered nationwide, including remote inaccessible rural regions. China has 202 of the 500 most powerful supercomputers in the world (America has only 143). In 2024 China was well recognised as the world leader on electric vehicles (EV). Its EVs are built-in and equipped with most up-to-date modern technologies in navigation and telecommunication systems.

Indeed, today China is a technological advanced and highly digitalised society. In 2023, digital economy contributed 66 percent of China's GDP[3] In 2024, China has rapidly expanded its digital economy into and integrated with other key economic areas such as agriculture, healthcare, and green energy. However, such rapid digitalisation in a short space of time would unavoidably give rise to negative side effect in the digitalisation process, as we saw in the case of Vietnam, and the price to pay for rapid digitalisation is cybercrime or technology-related crime.

3.2   Cybercrime in China and Its Criminal Justice System

Currently cybercrime problem in China is one of pressing issues for the Chinese Public Security authorities. With a population of 1.4 billion and most of them subscribing 5G mobile Internet networks, this would be a huge pool of target and an unresistable opportunity for potential cyber-perpetrator to pry their trade on to. More

---

[1] See CSET (Nov 27 2023) 'The Global Distribution of STEM Graduates: Which Countries Lead the Way' <The Global Distribution of STEM Graduates: Which Countries Lead the Way? | Center for Security and Emerging Technology (georgetown.edu)> . Accessed 27/08/2024.

[2] See China Association for Science and Technology <中国科学技术协会 (cast.org.cn)>. Accessed 29/08/2024.

[3] China Global News Television Network reported on 30 August 2024.

importantly, a large proportion of the Chinese netizens may not be highly technologically literate. According to 2023 figure,[4] over 70% of the older population between 60 and 70 aged were active netizens. 49 percent of the younger generations between 10 and 19 years were active netizens. Awareness of potential cyber-scammers of these two groups is likely to be low. According to Statista[5] report in 2017, the growth rate of cybercrime cases in China was at 57 percent and in 2020 it had reached 105%, of which majority were cyber-frauds. In the period from April 2021 to April 2022, cyberfrauds had increased to 394,000 cases and 634,000 arrests, almost double in just a year period.[6] Additionally, younger generation can fall victim to cybercrime, but they are also potential cyber-perpetrator. A Chinese's public prosecutors report suggested that 55% of the cybercriminal found guilty by the court of cyberfraud were between 18 and 28 of age[7] The China's cyberfraudster is getting younger. Chinese's youth generation may commit cybercrime unintentionally due to lack of awareness of the law and the actual situation they were in. There was also a lack of due diligence on the part of the college they were studying. In 2020 according to the Supreme People's Procuratorate report, 57 college students in Zhejiang province were offered telemarketing internships in corporations in various provinces in China. Though the whole process looks legitimate, the companies made telecom fraud rings that trick people into fake futures trading online, which neither the college nor the students were aware of this until when the police investigated the companies on suspicion of fraud and students were arrested in 2021. After lengthy police investigation, prosecutors concluded that the students have not wilfully intended to commit fraud and should not be prosecuted. Eventually in April 2022, authorities withdrew the cases and terminated the investigation.[8]

The fast pace of societal digitalisation drives cybercriminal today to other turfs, such as infringement of personal data. According to Chinese's prosecutor office report, there were filed charges against 280,000 individuals involved in cybercrimes in 2023, marking a 35.5% increase year-on-year. However, the prosecutor's office discovered through their investigations new types of cybercrimes using gimmicks such as the Metaverse, blockchain, and binary options platforms, with virtual currencies becoming breeding grounds for cybercrimes. While the more traditional type of cybercrimes such as online gambling, theft, pyramid schemes, and counterfeiting still existed, charges related to theft committed through the Internet increased by 22.7%, while charges related to online counterfeiting and sales of inferior goods surged by 85.7% from January to November. The prosecutor's office noted that there

---

[4] China Global News Television Network reported on 30 August 2024.

[5] See Statista <China: growth rate of cybercrime cases | Statista>. Accessed 29/08/2024.

[6] See Business and Technology (August 23, 2022) ' China's Cybercrime problem is growing' <China's cyber crime problem is growing – The China Project>. Accessed 29/08/2024.

[7] See Sina Finance (in Chinese) (1 August, 2022)
<司法大数据专题报告显示——涉信息网络犯罪案件量逐年上升，诈骗罪占比最高 新浪财经 新浪网 (sina.com.cn)>. Accessed 29/08/2024.

[8] See The Supreme People's Procuratorate of the People's Republic of China, March 20, 2023 'Prosecutions for telecoms, online fraudsters skyrocket' <Prosecutions for telecoms, online fraudsters skyrocket (spp.gov.cn)> . Accessed 30/08/2024.

is a growing trend on issues of insiders leaking citizens' personal information in industries such as finance, telecommunications, real estate, hotels, and labour intermediaries. In February 2024, over 7,300 individuals involved in the infringement of citizens' personal information were prosecuted.[9]

In addition, they also handled over 6,700 public interest litigation cases related to infringements of numerous citizens' personal information, failure of relevant entities to implement obligations related to combating telecom fraud, and illegal dissemination of false or harmful information during the period. Furthermore, cyber violence and online disruptive behaviours such as wanton insults, defamation, and privacy infringements severely violate the personal rights of others and seriously disrupt online order which are all on the increase too,[10] and in which formally charged 39 individuals with insult and defamation crimes

To give an example of online abuses, in June 2023, Hunan province's prosecution office investigated an online extortion case involving serving police officers. They found that between May 2021 to May 2022, Liu, Deputy Head of a local police station, abused his power and colluded with seven others to extort 357,000 Chinese yuan (US$50,000) from a man surnamed Zheng and others, when the prosecution office launched an investigation into Liu and the other seven individuals for suspected extortion. As a result, Liu and other five others were prosecuted. Liu was sentenced to 10 years and six months in prison, and the remaining five individuals received sentences ranging from one year and four months to seven years.[11]

Since 2024, we saw an increase of artificial intelligence (AI) generated cybercrime. Such perpetrators may not be based in China, but elsewhere in Southeast Asia. Through a connection in China, they committed cybercrime. Trend Micro[12] observed threat actors hiding malware in legitimate AI software, operating criminal LLM, thus enabling cybercriminals to trick generative AI bots into answering questions that go against their own pre-trained lawful policies—primarily for developing malware and social engineering lures. Also in the first quarter of 2024, cybercriminals have been ramping up deepfake offerings to carry out virtual kidnapping scams, fraud, and bypass KYC checks. Trojan malware has been developed to harvest biometric data to help with the latter—as some of victims were based in Hong Kong who were originally from Mainland China with a strong family connection inside China.

---

[9] See The Supreme People's Procuratorate of the People's Republic of China, February 23, 2024 'Crackdown intensifies on cybercrime' <u>Crackdown intensifies on cybercrime (spp.gov.cn)</u> . Accessed 30/08/2024.

[10] Op Cit.

[11] See The Supreme People's Procuratorate of the People's Republic of China, March 12, 2024 'procuratorates crack down on abuse' <u>Procuratorates crack down on abuses (spp.gov.cn)</u> . Accessed 30/08/2024.

[12] See Trend Micro (27 August, 2024) 'Alarming Surge in AI-Driven Cybercrime in H1 2024', <u>Alarming Surge in AI-Driven Cybercrime in H1 2024 | Media OutReach Newswire APAC (media-outreach.com)</u> . Accessed 30/08/2024.

3.3 Criminal Justice System

China is a vast country of huge land mass and substantial areas are rural such as farmland and countryside. It is understandable that there is a different or uneven application of criminal justice between urban modern cities and the rural country-side. Although criminal justice system is supposing to be nationwide in terms of application, but it is likely that there is more room for discretionary measure for the authorities in applying criminal justice in rural setting, where the community is far more likely to know each other personally than in the urban city. Due to such family ties, there is a relatively strong self-censorship already in place to restrain committing deviance or harmful acts to the community. However, in this section we will not distinguish rural or urban cities, and both rural and urban will be given the same treatment in the discussion.

Strictly speaking, China's criminal justice system consisted of three main institutions: the people's court system, the people's procuratorate system, and the public security system, 'the police'. Corresponding to this, Judicial structure in the Chinese broad sense does not only refer to courts, but also to procuratorates and public security organs too. This is because the security organ, 'the police', is one branch in the administrative system; while the other two branches are created by the people's congress and legally speaking, on an equal footing with the administrative branch. The presidents of courts and the procurator-generals of procuratorates are selected and appointed by the people's congresses on the same levels. The judges and procurators are selected and appointed by the standing committees of the respective people's congresses, and assistant judges and assistant procurators are appointed by the respective courts and procuratorates.

By putting together criminal justice system under one administrative organisational structure or branch, with a unified criminal justice system of which the National People's Congress (NPC) is the authority to oversee of the criminal justice and legal systems in China[13] it is the reason why that China's criminal justice systems have attracted a lot of critiques from outside legal scholar looking at from the western perspective. For example, the idea of check and balance and independent of the courts. Such an idea is embedded and is the core concept of separation of powers of the Western legal and justice systems. Therefore, the Chinese criminal justice system has a large gap to fill. However, when looking at the Chinese's criminal justice system from Chinese perspective and under the lens of Chinese societal context, especially in a historically sense, then we may start to see and understand why the difference existed. To start with, the Chinese criminal justice system is a very old system with more than 4000 years of history and only made some evolvements in recent years. The Chinese legal system was developed some 5000 years ago, when Confucius first introduced the concept of the importance of society over the individual, which doctrine survived through-out the imperial periods for thousands of years. Another

---

[13] In 1949 establishing its legal system, the Chinese's Communist Party created a unified system, with the NPC as the ultimate authority over all legal and governmental decision-making. This structure makes the NPC the overseer of the criminal justice and legal systems in China.

important socio-cultural concept was also the legacy of Confucius is deference to ancestors, leaders, and members of the elite class, putting these two concepts together, as Seay said:

> China depended on the beneficence of its leaders in the application of . Though laws may have been in writing, the application of them was left to the discretion of the individual leader - whether an emperor, a warlord, or a local governor. This discretionary approach became a "rule of the person," meaning that each person of authority could make a decision based on the prevailing beliefs, the most expedient choice, or the status of the person to be punished. This concept endured under Mao Zedong, despite an attempt to establish a classless or "uni-class society." The result was a wide range of vastly different punishments for what were often quite similar offenses.[14]

Not until 1911, we began to see the Chinese legal system being reshaped with a contemporary touch, which Seay has described neatly:

> As Sun Yat Sen led the transition from an imperialist and feudal state, to a nationalist state, and then later to a communist state. It was Mao Zedong who declared the creation of the PRC. in Tiananmen Square, Beijing, in 1949. It was also Mao Zedong who, in effect, abolished all laws in 1957. Following Mao's death in 1976, it was determined that the abolition of all laws was either a mistake and in 1979, Deng Xiao Ping authorized the creation of a written criminal code. A written constitution followed in 1982. These documents and rules were slowly implemented and enforced, first in Beijing, then in other major cities, and finally in the remote provinces and villages. It is these documents and laws, this evolution to a "rule of law.[15]

The evolution and development of Chinese's criminal justice system today is evolving around a socialist democracy—'socialism with Chinese characteristics'. The Chinese Constitution stated "*Our country is in the primary stage of socialism. The basic task before the nation is the concentration of efforts on socialist modernization construction in accordance with the theory of building socialism with Chinese characteristics*".[16]

The Chinese's criminal justice system we see in twenty-first century today still has a lot of works in progress and many gaps to be filled. Anyhow, we would like to find out whether the current form of the Chinese's justice system can withstand the tide of cybercrime in China which is next in turn, especially its response to cybercrime.

3.4   Response to Cybercrime

Digitalisation in the Chinese society has indescribably swift, from across the countryside to urban cities with fast speed 5G mobile network, enabling Chinese netizens to enjoy modern lifestyle of electronic commerce in selling and placing orders for

---

[14] Seay PA., 'LAW, CRIME, AND PUNISHMENT IN THE PEOPLE'S REPUBLIC OF CHINA: A COMPARATIVE INTRODUCTION TO THE CRIMINAL JUSTICE AND LEGAL SYSTEM OF THE PEOPLE'S REPUBLIC OF CHINA', IND. INT'L & COMP. L. REV, Vol. 9:1, p. 143.

[15] Op Cit. p,144.

[16] See XIANFA, preamble, para. 7, amended by XIANFA, amend. art. 3 (1993), also see Op Cit. p. 145.

products from all over the country with delivery satellite tracking system feeding live update into their mobile smart phone. In midst of this fast digitalisation process, China has experienced a painful explosion of cybercrime which the Chinese government has to respond. China therefore introduced ways to manage the wave of Internet connectivity within and outside of China, especially on cybercrime policing with an attempt to prevent harm done. A specific Internet-related law called 'Cybersecurity law' was introduced in November 2016 and implemented on June 2017. The Law principally covers three key areas (i) cybersecurity; (ii) national security; (iii) cyber-sovereignty. The legislative intent for this specific cyber regulatory was in the wake of 'Edward Snowden's[17] disclosure of the United States' government (including other five-eyes countries, United Kingdom, New Zealand, Australia, and Canada) secretly storing and using big data to analyse their own netizens' data, as well as spying on its netizen and netizens of other countries too. The objective of Chinese's cybersecurity law is to ensure that the cyber-space is secured; to safeguard cyberspace sovereignty, as well as to the safety of national security, social and public interests; the lawful rights and interests of the Chinese citizens, legal persons, and other lawful organisations; and in promoting the healthy development of the digitalisation of the economy and society in China. All in all, cybersecurity law's aspiration is practicality. However, the Western governments and critiques look at Chinese's cybersecurity law from another level in terms of restrictions of freedom of the netizen. But in fact, some of these Western countries have already enacted their own cybersecurity law long before the Chinese's law, and compared to the Chinese cybersecurity law, they are very much alike, if not hasher than the Chinese's, such as the United State Cybersecurity Act 2012. In any case, there were also many analysts with Western backgrounds who consider Chinese's cybersecurity law to be comparable to the European Union's (EU) General Data Protection Regulation (GDPR).[18]

Apart from regulatory response to cybercrime in China, there is also the introduction of specialist cyber police as early as in 2000s. Next, we will focus on China's police response to cybercrime.

3.5 Police Response

Back in 2006, China's police response to cybercrime was already in a novel way. The Shenzhen City's Police introduced a website Icon (as shown below) of virtual cop on their police reporting site. Once the Icon was clicked there were warnings

---

[17] The Guardian Newspaper, 8 June 2013 'Boundless Informant: NSA explainer', <Boundless Informant: NSA explainer – full document text I US news I theguardian.com> . Accessed 4/09/2024.

[18] See Bo, Q. U.; Changxu, H. U. O. (15 September 2020). "Privacy, National Security, and Internet Economy: An Explanation of China's Personal Information Protection Legislation". *Frontiers of Law in China*. **15** (3): 339–366. https://doi.org/10.3868/s050-009-020-0019-4. ProQuest 2,450,653,759.

on all sorts of cyber risks, together with other clear messages that the police were patrolling the Internet and the law was applied in cyberspace[19]

Nonetheless, it was the real police who would be answering the netizen enquiries, especially for cybercrime.

In a further drive to stamp the tide of cybercrime in China, the Ministry of Public Security announced in August 2015 that the cyber police teams would be working 24/7 across the country proactively to nip out illegal and harmful content on the Internet, to prevent cybercrime and as well as to publish reported cybercrime cases.[20] Besides the public security 'police' as specialist cyber police, China also has other primary law enforcement agencies in policing and investigating Internet-related crimes too, such as the Ministry of Industry and Information Technology (MIIT) and Internet Society of China (ISC). There are other more specific industry-related departments, such as for copyrights is the remit for the National Copyright Administration of PRC (NCAC) and for trademarks is the remit for the State Administration for Industry and Commerce (SAIC).

This clearly shows that China has developed various enforcement authorities trying to manage cybercrime efficiently. However, it is also clear that despite these policing (and laws), cybercrimes in China are on the rise, as more and more consumers log in and use the Internet for various activities. In this regard, in 2024 China also introduced artificial intelligence (AI) in both policing and investigating cybercrime[21] especially on cyber community policing in order to identify potential bank scam and to stop any fund before being transferred from the victims' accounts to the fraudster bank account online. In most cases the cyber police teams would dispatch a real police officer, either speaking in person to the potential victim on the phone or physically at the bank branch in a way to intersect before the potential victim actually transfers money from her/his bank account to the fraudster's bank account (most of the time these perpetrators' bank accounts are based overseas). In the twenty-first century this was called 'influence policing' model and this kind of policing method has already

---

[19] See Shanghai Daily Newspaper, (January 5, 2006) 'Cyber Police to Guard All Shenzhen Websites'.

[20] See The State Council, The People's Republic of China, Xinhua, (August 12, 2015) ' China beefs up cyber police force', <China beefs up cyber police force (www.gov.cn)> . Accessed 4/09/2024.

[21] See Zhang L. Y. (in Chinese), (July 2024) Artificial Intelligence Community Safety Novel Solutions, in Chinese Sociology Association Annual Congress Proceedings, P.298.

been used in policing in various Western societies, not just in China, for example in Scotland, Scottish Police used it for the prevention of Internet-related crimes.[22]

3.6 Summary

In short, although China, like Vietnam, is a socialist country, Chinese socialism state is carrying a strong flavour of Chinese capitalist characteristics. Unlike its neighbour Vietnam, China is a huge country with 1.4 billion of population and geographic land mass covering no less than 9.6 million square kilometres. If one is to take a flight from the East side of China to the Northwest side, it takes approximately 7–8 h of flight time. Therefore, we must not judge China and Vietnam on an equal measure. Additionally, China is a super power state both economically and politically in the Asia Pacific region and if not globally. China is leading in technological advancements and innovations, and because of this, China has projected itself onto the globally stage as the global player for manufacturing and application of latest information communication and technologies (ICTs) at the heart of the Chinese society. However, as China is rapidly transforming itself (as fast as its high-speed railways) into digitalised society, China, similar to Vietnam in some senses, is struggling to get a grip with the fast rate of growth of cybercrime. However, it appears that China has the capability, determination, and financial resources to fight cybercrime, most importantly, China as the leader in the technological pack globallycan innovate by applying the latest technology to fight cybercrime, such as AI in policing. Even though Vietnam and China were using crime control in policing cybercrime, China is way ahead and better by deploying its centralised policing effectively. In ending this Volume 1, we'll offer a short conclusion in the next chapter.

---

[22] See Collier B. et al., (August 2023) 'INFLUENCE POLICING: Strategic communications, digital nudges, and behaviour change marketing in Scottish and UK preventative policing' in Scottish Institute for Policing Research, Future of Policing Report Series.

Chapter 4
Conclusions: Minimising and Managing Cybercrime Risks—People's Republic of China and Vietnam

As we see in previous chapters, China and Vietnam are experiencing rapid digitalisation in their societies. The COVID-19 global pandemic[1] had directly and indirectly pushed most of us to go online for remote working, shopping, and leisure too. Technologisation and datafication in both countries are most likely much deeper than that and most likely important public policy of economic development and planning is involved too. Since both China and Vietnam are going through rapid societal informatisation, a discussion will be given on how both countries attempt to minimise and manage their cybercrime risks, and on the harm done to their citizens and the society.

4.1 Technologisation and Datafication

Technologisation and datafication in Asia have been particularly visible and rapid in recent decades, as we saw both Vietnam and PRC in Volume 1. In fact, such a trend is global, albeit some countries are faster than others. In the case of PRC, technologisation and datafication have been remarkable, transforming China from a very poor and underdeveloped country into a global tech. powerhouse in the twenty-first century. Chinese government plays a key role here, and it is no accident that while China rises to the global stage, public policy at the highest level plays a deciding role in fostering technological growth as we see today in Chinese society. For example, two decades ago China was already planning for 'Made in China 2025' initiative[2] (as a part of

---

[1] See Slay J., 'Pandemics and illegal Manipulation of Digital Technologies: Examining Cause and Effect in a Time of Covid-19', in R. G. Smith, R. Sarre, L. Y-C. Chang, L. Y-C. Lau (Eds.), (2023) *Cybercrime in the pandemic digital age and beyond. Palgrave studies in cybercrime and cybersecurity.* Palgrave Macmillan: London, p. 17.

[2] This plan is to independence from foreign supplies, and to encourages increased production in technological products and services, with its computes and chips industries central to the plan.

L. Yiu-Chung Lau, *Cybercrime in Asia*, SpringerBriefs in Cybersecurity, https://doi.org/10.1007/978-3-031-80213-3_4

the thirteenth and fourteenth five-year plans) as its key national strategic plan and industrial policy. China invested US\$300 billion into achieving the industrial plan and further invested US\$1.4 trillion during the global COVID-19 pandemic period. The Chinese leadership sees China's economic success as a direct link to its industrial policy planning, particularly for the Fourth Industrial Revolution.[3] Central to the 2025 initiative is to identify key technologies, such as AI, 5G (and 6G), aerospace, computers and chips, electric vehicles, and biotech, thereby indigenizing those technologies with the help of national champions, securing market share domestically within China, and ultimately capturing foreign markets globally. Because of this 2025 initiative, China invests heavily in research and development in key areas such as AI, robotics, renewable energy, and space technology.

To further support the 2025 initiative, China has developed extensive infrastructure to facilitate the country's technological growth. A good example in point is the high-speed train networks (largest in the world, in terms of coverage in distance today), 5G mobile networks, and largest number of smart cities globally.

All these, consequently, led China into the world's largest digital economy (66 percent of GDP today). Today datafication in Chinese society is deep and wide. A few technology champions have helped and driven datafication integrated into ordinary life inside the Chinese society. For example, technology companies such as Huawei, Tencent, Baidu, and Alibaba help push the boundaries in AI and 5G technologies. Today, nearly a billion netizens adopting and using these technologies champion services and products, therefore fueling the innovations in e-commerce, social media, on the demand services, as well as mobile payment systems, such as fintech, i.e. payment with QR code.

Chinese netizens are ready to accept and adopt social application in their smart mobile phone, helping what once regarded as fictional stories of people living in virtual reality worlds as depicted in popular novels are now rapidly transitioning into the real world. One good example is the Black Myth WuKong role-play gaming released on 20 August 2024, another example is the release of the Chinese's cartoon movie 'Ne Zha 2' on late January 2025, this cartoon movie was assisted and made with advance Chinese home-grown AI tecnologies proof a winner globally. Chinese consumers are moving more and more of their social interactions and leisure activities into virtual domains. Virtual and physical social activities are also merging, as offline social interactions are increasingly orchestrated through virtual communities, where offline meetups more frequently spill over from communities and connections initially formed online. For example, in Shanghai, an app-linked bicycle community has expanded its activities into other areas of social life. On online games streaming sites such as Huya and DouYu, people brought together by the love of a particular title, genre, or livestream host are taking chatroom interactions into the real world, organising yoga classes, and running clubs. This trend led Tencent to build a portfolio that merges gaming, e-commerce, and social, creating a decentralised, competitive, and creator-friendly ecosystem—a metaverse, this would be impacting the physical world.

---

[3] This is the term use for twenty-first centuries rapid technological advancement in a society.

Whereas in the case of Vietnam, societal datafication is also rapid. This is because the Vietnamese government is emulating China in their drive for technologisation of their economic development, the Vietnamese government economic development policy aims to be a new powerhouse of the Southeast Asia region in the digital economy through its 2030 initiative. The key aim for such an initiative is to reach US$220 billion in monetary terms by 2030 and contribute 30% of their GDP from digital economy. At the same time, Vietnam, like China, is aiming to establish smart cities, but in specific key economic zones, thereby creating a different set of regulatory rules.

These special economic zones are trying to attract foreign direct investment (FDI). In some way Vietnam has succeeded with its friendly public policies, including tax incentives. A number of major technology companies, such as Apple and Samsung, are making inroad into Vietnam and establishing a manufacturing base there. This is partly to do with geopolitics at play between China and the US (this point we will be discussing further later in Volume 3) as US is decoupling from China. These technology companies want to diversify their investment and eventually benefit Vietnam. FDI continues to flow in for development and manufacturing by companies such as Microsoft, Sony, Pegatron, Nokia, Panasonic, Intel, and Canon. Vietnam is now also a regional hub for research and development (R&D) outsourcing for Cisco, Alcatel-Lucent, Toshiba, Hitachi, and Jupiter Networks. Vietnam is on its way to becoming a key semiconductor player in the region and an increasingly critical link in the global supply chain too. More importantly, Vietnam is also aiming to be among the top rank in Industrial Revolution 4.0 in Southeast Asia.

All these factors have greatly helped Vietnam's technologisation and datafication in their society in a short space of time. Vietnam's high Internet penetration rate of 72% of the population and a high number of tech start-up established by the youth population[4] are the key factors for datafication in Vietnamese society today, since these young generations are finding occupations in hi-tech exciting and they are also embracing tech for leisure too. The implication for societal technologisation and datafication is echoed in their cybercrime rates. How to mitigate the crime risks is one of key tasks for both the Chinese government and Vietnamese government.

4.2 Minimising and Managing Cybercrime Risks

4.2.1 The Chinese's Style

To minimise and manage cybercrime, China has a comprehensive approach to its policing. Regulatory framework is relatively well developed as we saw in Chapter Three, and at the functional level Chinese police is innovative, using technology

---

[4] Vietnam Recorded the highest proportion of young people in the country's history with 20.4 million youth, aged 10-24, accounted for 21 percent of the total population, <Vietnam - Healthy Cities for Adolescents (healthy-cities.org)>. Accessed 9/09/2024.

to fight technology—AI application is probably one of Chinese policing priorities. On national level cybercrime is on top of the top policing priorities for the police, and cybercrime is one of their annual key performance indicators (KPI) for all police district commissioners in China.[5] One reason is that cybercrime can be a disrupting force for social harmony and harm digital economy growth (cybercrime is going against capitalist with socialist Chinese character). Nonetheless, cybercrime is evolving rapidly to artificial intelligence too[6] and the public security agencies are also resolving to AI as well in their policing. This is because China today has the financial resources as well as capability to do so. Law enforcement training is relatively comprehensive and multifaceted. China's police training programmes are designed to develop specialised public security personnel at different levels in universities (today 60% of serving police officers in China are university graduates), colleges, and schools; to train police officers according to their different grades; and to produce specialised personnel of all ranks. The goals of higher institutions are to foster specialisation, including cybercrime policing and high-level public security personnel, and to train leading police officers above the rank of directors of county public security bureaus or sections. China also sets up vocational schools with 2-year or 3-year programmes run by provinces, autonomous regions, and municipalities. The Public Security Administrative Cadres College admits applicants for public security organisations who have graduated from middle schools or have an equivalent educational level. The People's Police School is a kind of specialised secondary school that admits graduates from senior middle schools. Furthermore, China has established international cooperation with international organisation, such as the INTERPOL, and China has created a special police department called 'National Central Bureau (NCB)' to cooperate with the INTERPOL and regularly takes part in global INTERPOL led police operations, supporting cooperative efforts to tackle terrorism, cybercrime, drug trafficking, intellectual property crime, and environmental crime[7] However, under the current climate of ongoing geopolitics (unless this changes any time soon) between China and the US, this could be a challenge for any coordinated global effect to fight cybercrime, as both China and the US would likely look and do things inwardly, instead of from a global perspective. This is greatly dampening the effectiveness and efficiency of global policing on cybercrime, since the Internet is boundless, and it does not recognise any border.

---

[5] This is confirmed to the author by a district police commissioner, on early July 2024 as he went on to a police district visit and exchange at Harbin City, after a brief tour of the police district and station, in which during in the 3 h long exchanging view and discussion session, that the district police commissioner clearly indicated to us that cybercrime nationally is the most important 'number 1' crime for them to deal with as police officer, particularly on prevention of before the cybercrime is happening and to pursue the perpetrator both in China and aboard, in order to prosecuting the cybercriminal in court.

[6] See Euronews, (8 May 2024) 'The cost of cybercrime around the world is estimated to reach 11.2 trillion euros a year by 2025', <Artificial intelligence fuelling global surge in cybercrime | Euronews > . Accessed 5/09/2024.

[7] See National Central Bureau – INTERPOL <China (interpol.int)> . Accessed 9/09/2024.

4.3   The Vietnamese Style

Vietnamese policing cybercrime is quite different from the Chinese, and Vietnamese policing priority is likely to be different, so in turn the way of managing cybercrime risks is likely to be different, as Lau argues "*…each country policing is different due their socio-economics, socio-politics, police organisation and structure, so their policing priorities were difference…..*[8]" Nevertheless, the Vietnamese government has established a regulatory framework, albeit it is quite underdeveloped. The Vietnamese Cybersecurity law of 2018 is a key piece of law dealing with national security on cyberspace, apart from the early 2015 Criminal Code addressing offences such as illegal access, computer fraud, misuse of computer devices, and interception. Cybersecurity law 2018 is the latest key legislation that deals with the most pressing issues of national security on cyberspace, including cybercrime, and it officially took effect on 1st January 2019. This law covers a wide area. Article 8 specified acts against spreading false information online, Article 26.3 deals with data storage and that foreign enterprise must store their data inside Vietnam, together with Article 16 all enterprises on request from the authority must provide access to data stored in their computer systems and halt the computer systems from operating, as well as reporting any crime to the authority within 24 h, if any enterprise suspects cybercrime is happening in their systems. Whereas Article 17 is looking at spying online activities, so that eavesdropping online conversations is cyber espionage, Article 29 of the Cybersecurity Law is to protect children from harm of the Internet.

Enacting law is one thing but actually dealing with cybercrime is another. In practice policing cybercrime in Vietnam is a relatively new skill for the police, though the Vietnamese police have been receiving training from various different international organisations, including the United Nation Office on Drugs and Crime (UNODC) to better handle cybercrime. Particular key skill sets such as how to collect and preserve electronic evidence and manage investigation procedures should be taken up on a wider scale. Challenges remain for the police in Vietnam, due to financial resources and capability restraints. In recent years, Vietnam has experienced a significant increase in cybercrimes within and from the outside and is considered one of the emerging hotspots for cybercrime with Vietnamese hacker groups involved in transnational cyberattacks as we saw in Chapter Two. This is not helpful for Vietnamese law enforcement agencies as the policing resources are already finite, plus cybercrime is likely not their policing priority. Even though, Vietnam is proactively improving its cybercrime policing capabilities, by participating and collaborating with international organisation, i.e. UNODC, but it is still working in progress for the law enforcement agencies on minimising and managing cybercrime to create a safer cyberspace for its citizens.

---

[8] See Laurie Yiu Chung Lau, (2005) 'Governance in the Digital Age: Policing the Internet in Hong Kong', in Broadhurst R. Grabosky P. (edited) Cyber-crime: the Challenge in Asia, Hong Kong: HKU Press, P. 89.

4.4 Closing Remarks

In short, it is clear that to combat or to minimise and manage cybercrime, the nation states must utilise, on and through a combination of comprehensive legal frameworks (best to evolving with technology, as the technology evolves), a well-funded law enforcement agencies with capability and training, as well as international cooperation together with global policing networks. However, such international cooperation also depends on the direction of the windblown in terms of geopolitics. Perhaps, in addition to the above, the awareness of the respective country's netizens on data security, especially on what not to do and what to do when clicking on any mobile application (think before you click scenarios here, namely this is to do with the task of public awareness and education). Lastly the level of willingness of the private sectors, including individual ownership and small and medium enterprises (SMEs), as well as large corporations in cooperating with the legitimate investigating authorities on reporting and providing or allowing access to data for the investigator to complete the prosecution process, to bring the perpetrator to face justice in court it is important too.

Bibliography

Books and Journals

1. Bo, Q. U.; Changxu, H. U. O. (15 September 2020). "Privacy, National Security, and Internet Economy: An Explanation of China's Personal Information Protection Legislation". *Frontiers of Law in China.* **15** (3):
2. Collier B et al, (August 2023) 'INFLUENCE POLICING: Strategic communications, digital nudges, and behaviour change marketing in Scottish and UK preventative policing' in Scottish Institute for Policing Research, Future of Policing Report Series.
3. Dang Thi Viet Duc et al, 'Measuring the digital economy in Vietnam', in Telecommunication Policy, Vol. 48, Issue 3, April 2024, https://doi.org/10.1016/j.telpol.2023.102683.
4. International Monetary Fund Working Paper, Longmei Zhang, Sally Chen (2019) 'China's Digital Economy: opportunities and risks'.
5. Laurie Yiu Chung Lau, (2005) 'Governance in the Digital Age: Policing the Internet in Hong Kong', in Broadhurst R. Grabosky P. (edited) Cyber-crime: the Challenge in Asia, Hong Kong: HKU Press, P. 89.
6. Le, L.C., Hoang, Y.H., Bui, H.T. *et al.* Understanding Causes for Wrongful Convictions in Vietnam: a View from the Top and the Bottom of the Iceberg. *Asian J Criminol* **17** (Suppl 1), 55–73 (2022).
7. Luong et al 'Understanding Cybercrimes in Vietnam: From Leading-Point Provisions to Legislation System and Law Enforcement' K. Jaishankar : Open Access, International Journal of Cyber Criminology July- December 2019 Vol. 13.
8. Lusthaus, J 'Modelling cybercrime development: The case of Vietnam' in 2020 (eds.) LeuFeldt R Holt T J, The Human Factor of Cybercrime, London: Routledge, p. 246–55.
9. Slay J., 'Pandemics and illegal Manipulation of Digital Technologies: Examining Cause and Effect in a Time of Covid-19', in R. G. Smith, R. Sarre, L. Y-C. Chang, L. Y-C. Lau (Eds.), (2023) Cybercrime in the pandemic digital age and beyond. Palgrave studies in cybercrime and cybersecurity. Palgrave Macmillan: London
10. Seay PA., 'LAW, CRIME, AND PUNISHMENT IN THE PEOPLE'S REPUBLIC OF CHINA: A COMPARATIVE INTRODUCTION TO THE CRIMINAL JUSTICE AND LEGAL SYSTEM OF THE PEOPLE'S REPUBLIC OF CHINA', IND. INT'L & COMP. L. REV, Vol. 9:1, p.143.
11. Smith, R. Sarre, L. Y-C. Chang, L. Y-C. Lau (Eds.), (2023) *Cybercrime in the pandemic digital age and beyond. Palgrave studies in cybercrime and cybersecurity.* Palgrave Macmillan: London
12. Zhang L Y (in Chinese), (July 2024) Artificial Intelligence Community Safety Novel Solutions, in Chinese Sociology Association Annual Congress Proceedings

© The Editor(s) (if applicable) and The Author(s), under exclusive license to Springer Nature Switzerland AG 2025
L. Yiu-Chung Lau, *Cybercrime in Asia*,
SpringerBriefs in Cybersecurity, https://doi.org/10.1007/978-3-031-80213-3

Online

13. Bank Info Security, Jayant Chakravarti (April 9 2024) 'Vietnam Struggling to Contain Growing Cybercrime Ecosystem', <Vietnam Struggling to Contain Growing Cybercrime Ecosystem (bankinfosecurity.asia)>.
14. Bangkok Post, <https://www.bangkokpost.com/business/general/2813479/fortinet-points-to-surge-in-value-of-cybercrime-by-2025>.
15. The Bangkok Post <https://www.bangkokpost.com/business/2017999/digital-hub-status-within-reach>.
16. BBC Online Report 'This is the first time that cloud-based face verification has been used to secure the identity of people who are using a national digital identity scheme' <https://www.bbc.com/news/business-54266602>
17. BBC Online (28 August 2024) 'World's police in technological arms race with Nigerian mafia' <Nigeria's Black Axe mafia dealt 'big blow' by Interpol (bbc.com)>.
18. Business and Technology (August 23, 2022) ' China's Cybercrime problem is growing' <China's cyber crime problem is growing – The China Project>.
19. Bloomberg: https://www.bloomberg.com/news/articles/2019-12-22/vietnam-linked-hacking-group-targets-toyota-other-companies.
20. Global Cyberattacks Continue to Rise with Africa and APAC Suffering Most - Check Point Blog .
21. China Association for Science and Technology <中国科学技术协会 (cast.org.cn)>.
22. The People's Republic of China, State Council, Press Release in English, <Digital sector roadmap to aid recovery (www.gov.cn)>
23. The People's Republic of China, State Council, Press Release in English, <Digital economy expands in scale, demonstrating enormous potential (www.gov.cn)>.
24. The Supreme People's Procuratorate of the People's Republic of China, February 23, 2024 'Crackdown intensifies on cybercrime' <Crackdown intensifies on cybercrime (spp.gov.cn)>.
25. The Supreme People's Procuratorate of the People's Republic of China, March 12, 2024 'procuratorates crack down on abuse' <Procuratorates crack down on abuses (spp.gov.cn)>.
26. The Supreme People's Procuratorate of the People's Republic of China, March 20, 2023 'Prosecutions for telecoms, online fraudsters skyrocket' <Prosecutions for telecoms, online fraudsters skyrocket (spp.gov.cn)>.
27. CSET (Nov 27 2023) 'The Global Distribution of STEM Graduates: Which Countries Lead the Way' <The Global Distribution of STEM Graduates: Which Countries Lead the Way? | Center for Security and Emerging Technology (georgetown.edu)>.
28. Euronews, (8 May 2024) 'The cost of cybercrime around the world is estimated to reach 11.2 trillion euros a year by 2025', <Artificial intelligence fuelling global surge in cybercrime | Euronews>.National Central Bureau – INTERPOL <China (interpol.int)>.
29. International Monetary Fund Estimate for 2019 http://www.internetlivestats.com/internet-users-by-country/.
30. Krasia https://kr-asia.com/vietnam-suffers-the-most-southeast-asia-offline-cyber-attacks-q2-2019/.
31. GlobalEconomy : https://www.theglobaleconomy.com/Vietnam/Literacy_rate/ The Guardian Newspaper, 8 June 2013 'Boundless Informant: NSA explainer', <Boundless Informant: NSA explainer – full document text | US news | theguardian.com>.
32. QSWowNews:https://qswownews.com/university-education-may-lead-worse-job-prospects-vietnam/ .
33. SCMP <https://www.scmp.com/news/hong-kong/community/article/2139170/least-2-million-internet-users-hong-kong-were-hit>.
34. SCMP Newspaper Editorial 'Hong Kong's new cybercrime team can help turn rising tide of scams' 15 July 2023, <Opinion | Hong Kong's new cybercrime team can help turn rising tide of scams | South China Morning Post (scmp.com)>.
35. Statista <Chart: Cybercrime Expected To Skyrocket in Coming Years | Statista>.

36. Statista <China: growth rate of cybercrime cases I Statista>.
37. Shanghai Daily Newspaper, (January 5, 2006) 'Cyber Police to Guard All Shenzhen Websites'.See The State Council, The People's Republic of China, Xinhua, (August 12, 2015) ' China beefs up cyber police force', <China beefs up cyber police force (www.gov.cn)>
38. Sina Finance (in Chinese) (1 August, 2022) < 司法大数据专题报告显示—— 涉信息网络犯罪案件量逐年上升，诈骗罪占比最高 新浪财经 新浪网 (sina.com.cn)gt;.
39. Transparent International: https://towardstransparency.vn/en/vietnam-cpi-2019-score-is-up-but-corruption-remains-serious/.
40. Trend Micro (27 August, 2024) 'Alarming Surge in AI-Driven Cybercrime in H1 2024', <Alarming Surge in AI-Driven Cybercrime in H1 2024 I Media OutReach Newswire APAC (media-outreach.com)>.
41. Vietnam+ Sci-Tech April 4, 2024 'Over 20 million cyberattack warnings detected in Q1', <Over 20 million cyberattack warnings detected in Q1 I Vietnam+ (VietnamPlus)>.
42. Vietnam Recorded the highest proportion of young people in the country's history with 20.4 million youth, aged 10–24, accounted for 21 percent of the total population, <Vietnam - Healthy Cities for Adolescents (healthy-cities.org)>.
43. World Economic Forum <Asia Pacific region is the new ground zero for cybercrime I World Economic Forum (weforum.org)>.

Others

44. China Global News Television Network.
45. XIANFA, preamble, para. 7, amended by XIANFA, amend. art. 3 (1993).

Index